101 Ways to Live Cleaner and Greener for Free

Written by
Anna Pitt

Illustrated by
Toni Lebusque

For my children, their children
and their children's children

Published by: Green Lanes Publishing

A CIP catalogue record for this book is available from the British Library

ISBN 978-0-9574637-0-7

I am quite a 'green teen' but this book still gave me so many simple and fun tips to help our planet. I am going to try and do them all! I hope more people my age read this, whether they are already quite green or not at all. The maths and science parts were also very interesting and eye-opening!

I also have a tip: I use old cardboard from boxes like cereal and then old magazine cuttings to make birthday and Christmas cards - saves a lot of money and people love them.

Isabella di Stefano, sixth form student, age 17

This book is proper good. I actually laughed out loud twice. The intro was really funny and the pictures and the tips people sent in were fun and a bit different.

Anna Harris, age 16

Fascinating. The maths part really made me think about how much we wasted and altogether a very interesting book. Definitely worth a read.

Milly Foxcroft, age 15

I love it...this should be supplied by the Government for every household! Anna Pitt has done a fab job. The structure is brilliant and makes for easy reference when you want to read the relevant sections. #impressed

Lisa Cherry, Author of Soul Journey

This is actually really interesting. I keep wanting to turn over the page to see what's next. I've come to some tips and thought, 'She doesn't do that.' And then I realised, 'Oh wait, she actually does!'

Jennifer Pitt, age 16

Going green doesn't have to cost the earth. In fact, it doesn't have to cost anything at all. Here's how...

Contents

Introduction

What does being green really mean?

The term 'green', apart from being a colour, has been used to mean many things. It has been used to describe people who are new to something. It can be used to imply jealousy or envy. But mostly these days 'being green' is about caring for our planet.

If you have opened this book and are reading these words then I don't need to persuade you that it's important to be green. You've already made that decision, haven't you, or surely you wouldn't even have opened the book? Unless, of course, you just like the pictures.

You can be green in so many different ways. You can invest your savings in solar panels to put on your roof. You can buy eco-friendly holidays. You can eat organic food or you can change to a brand of eco-friendly washing-up liquid. But these things cost money and often they cost more money than the non-green option. And there's something about that state of affairs that isn't quite right, don't you think?

I'm never entirely sure whether an eco-friendly product is more expensive than its less eco-friendly counterpart because the constituent parts or ingredients are more expensive to produce. Or are they more expensive as people will pay a premium for being green because they think it is important and so manufacturers take advantage of that fact?

I'm glad the electricity I'm using to type this book is from entirely renewable sources, from the solar panels on my roof and from my 100% renewable electricity supplier, Good Energy. I prefer to drink orange juice that is delivered to my door by my milkman in reusable glass bottles. But this book is not about buying solar panels, organic vegetables or eco washing-up liquid. It is about things that you can do that are entirely free.

If that is the 'green' part what about the 'clean' part?

Well I'm not suggesting that any of us need to wash more. In fact today, we wash ourselves, our clothes and everything around us way more than ever before. Historians tell us that years ago people used to pong. I guess most of us can pong a bit now after some serious exercise, but as a general rule we soon get ourselves smelling sweet again once we've finished. So what I'm really talking about when I use the word 'clean' is about keeping our planet clean. I'm talking about living in a way that doesn't leave behind a whole load of mess in the form of waste and pollution.

So really I use the terms 'cleaner' and 'greener' to mean the same thing. Maybe some people like to call it 'clean' and others like to call it 'green'.

This book is a collection of 'eco-tips', many of which were sent in by people like you and me, who know that we need to take care of the environment and who have made the decision to go clean and green, in whatever small ways they can. The one rule was that people could do all these things for free.

While collecting the tips so many people have said to me what a great idea they think this is, but equally lots of people have then been dismayed that they can't think of a tip that is entirely free. It is worrying that in some ways this book has served to highlight that so many people are put off by the fact that the 'green economy' is seen as a trendy way of making a few quid.

Most people who read this book will probably already be doing some of the things it mentions. Some who read this book might be doing all of these things and more. You may be doing some or all of these things in your family. But if you are also doing things that aren't mentioned, then please do send in those ideas for new editions, for workshops and for the blog.

If you are not doing any of these things, then start slowly. Pick just one or two ideas and see if you can make them work for you today. Don't forget, they won't cost you a penny so you have nothing to lose. In fact many of the tips will save you money so you have plenty to win. How to choose where to start? It really doesn't matter. You could pick the one with your favourite illustration, the one that seems so ridiculous it makes you laugh, the one that seems so obvious you can't believe you didn't think of it yourself. Or you could start at the beginning.

When you have finished reading, please take a look at the section at the end on what you can do with this book. I hope it will make you smile!

Reducing Food Waste

The Science

What was wrong with the old way of dealing with food waste?

We used to think it was okay to throw away food waste, along with other kinds of household rubbish, into the dustbin. That's what everyone did. The dustbin was collected and taken to a landfill site, where it would be compacted and buried and we all just looked the other way at these ugly places and held our noses if they smelled bad.

But we mainly did this because we didn't know what else to do and we didn't realise that we were damaging the environment. These days we know a lot more about what happens to our rubbish when it is in a landfill site. We are aware of the problems we are creating for the planet.

We know that food waste rots after a while. Most of us have probably seen it even in our own fridges! It is the microbes in the food that break it down. The problem with food waste decaying in landfill sites is that the broken down organic matter disperses into water (e.g. rainwater) and other liquids in the waste. This liquid mixture is known as 'leachate'. The problem is, this leachate can become toxic. If we don't want this to cause pollution, then it has to be collected and treated to remove the harmful substances.

Apart from this potentially toxic leachate liquid, the rotting organic matter in landfill also produces gases, such as methane and carbon dioxide, known as greenhouse gases. Landfill gas contains around 40-60% methane, with the remainder being mainly carbon dioxide.[1] There are

---

1 Ewell, M. Primer on Landfill Gas, www.energyjustice.net/lfg/

also traces of nitrogen, oxygen, water vapour and lots of contaminants, some of which can even be radioactive. You may have heard of chloroform or carbon tetrachloride, which are both poisonous substances that can be found in landfill gas.

So the old way of dealing with rubbish basically has three problems:

- *Firstly, it requires space for an unsightly and smelly mass of stuff. We don't want to keep using more and more space for this. We will eventually run out of possible places to dump our waste.*

- *Secondly, it produces leachate – which can become toxic and has to be managed so that it doesn't get into our rivers and pollute them.*

- *Thirdly, it produces greenhouse gases such as methane and carbon dioxide, as well as some toxic gases.*

What is the new way of dealing with food waste and why is it better?

In many parts of the UK today, local councils provide a special bin for food waste. This bin is usually a tightly sealed container so that rodents can't get in and smells can't get out. The food waste from these bins is picked up by the local council's collection service and taken to a food waste processing plant. These food waste processing plants use either composting or something called 'anaerobic digestion'.

Anaerobic digestion (let's call it AD) is a way of breaking down the food to convert it into fertiliser for farmland and electricity that can be used to power homes, schools and businesses.

'Anaerobic' means without oxygen. In an anaerobic system, the waste is put into sealed containers so that air can't get in. Microorganisms break down the organic matter (i.e. the waste) into biogas (a mixture of methane and carbon dioxide) and digestate (a nitrogen-rich fertiliser). The biogas can be used to produce electricity and the digestate can be used to fertilise farmland.

But hang on a minute, doesn't that sound like the same gases produced in landfill? So, why is biogas considered different to landfill gas? How is it any better?

Landfill gas and biogas from AD do contain a similar mix of gases. Like biogas, landfill gas can be captured and used to produce heat and energy. The quality of the gases depends on what is put in. Landfills have a mixed input and it is the mixture of source material that can give rise to the toxic elements of the gas. For example, batteries, fluorescent light bulbs, electric switches, thermometers and paints are particularly problematic, because they contain mercury. AD plants tend to have highly controlled input which makes them easier to predict and manage. Both sources of gas contain toxins which need to be removed and processed into safe substances. Because what goes in to an AD plant is controlled, the processing is simpler and more cost-effective. And because of the way an AD plant is designed, the gases are captured and used for heat and energy.

Then there's the benefits of using what's left as fertiliser. You probably already know that farmers need to add nitrogen and phosphates back into the soil. Some fertilisers are derived from inorganic sources and these are non-renewable. Digestate is a renewable source of organic fertiliser so it can offer carbon savings associated with mining, transport and production of chemical fertilisers.

So the new way of dealing with food waste basically has three advantages:

- *It reduces the pressure on landfill spaces.*
- *The gases are used to create heat and energy.*
- *The remaining digestate provides a source of fertiliser.*

That sounds great, so where's the problem? Well, the only problem really is us. You know when you are used to doing one thing and then someone asks you to do it differently, that's not always easy at first, is it? The thing is though, it makes a huge difference when we do make the effort to deal with our food waste in this new way.

But what is the best way of dealing with food waste?

To have less of it!

If you are constantly filling your food waste bin every week then maybe your family is buying or cooking too much food. You could save quite a bit of money by looking at what is being thrown away on a weekly basis. Why not try out Tip 3 and have a go at planning a week's meals before doing the food shop? Think about ways of using up the leftovers. Or perhaps suggest smaller servings if your family are not finishing the food on their plates. You could always have fruit or cake if any of you are still hungry!

Most of the food in a food waste bin is money being thrown away, so it is well worth trying out ways to reduce this waste.

The Maths

UK households produce around 8,000,000 tonnes of food and drink waste every year[2]. Every tonne of food and drink that is consumed rather than sent to landfill represents a saving of 4.2 tonnes of CO_2 equivalent emissions (CO_2e). By comparison, every tonne of food waste sent to anaerobic digestion instead of to landfill represents a saving of 0.5 tonnes of CO_2e emissions.[3]

4.2 tonnes - 0.5 tonnes = 3.7 tonnes

So every tonne of food we eat rather than waste represents a carbon saving of at least 3.7 tonnes. That's about the weight of a female elephant.

But a tonne of food waste is a lot, isn't it! Presumably that tonne of food waste is thrown out by a lot of people. Or is it? There are 26,300,000 households in the UK.

8,000,000 tonnes ÷ 26,300,000 = 0.3041825095057

Wow! That's nearly a third of a tonne of waste per household in one year. So, in just over three years, your household alone will have generated (on average) a tonne of food waste. Or looked at another way: your household and your two next-door neighbours will, likely, throw out a tonne of food just this year. Does that shock you?

2 DEFRA gives the 2009 figure for food and drink waste as approximately 8.3 million tonnes. http://www.defra.gov.uk/environment/waste/

3 http://www.defra.gov.uk/publications/files/anaerobic-digestion-strat-action-plan.pdf

It's estimated that UK households waste a quarter of the food they buy unnecessarily. If your family spends £60 per week on food, then by cutting out avoidable food waste for a year you could save:

$$£60 \div 4 = £15$$

$$£15 \times 52 \text{ weeks} = £780 \text{ per year.}$$

If your family spends £100 per week, then that saving becomes:

$$£100 \div 4 = £25$$

$$£25 \times 52 \text{ weeks} = £1300 \text{ per year.}$$

Not to be sniffed at, hey?

One of my favourite tips is number 11. Shall we take a look at the sums behind that one?

Lots of people who go out to work buy their sandwiches for lunch every day. Taking a packed lunch of yesterday's leftover dinner costs £0.00 compared to, for example, a 'Meal Deal' (which might be a packet of sandwiches, a bag of fruit and a drink) for £2.99. Imagine that someone takes in leftovers for lunch once a week for a whole year. We'll give them five weeks holiday. We know they normally spend £2.99 on their lunch. Let's do the maths...

$$52 - 5 = 47 \text{ (weeks worked in a year)}$$

$$£2.99 \times 47 = £140.53 \text{ (annual saving)}$$

Now, that would buy a few treats at Christmas, wouldn't it?

The great thing about these tips that help to reduce food waste is that they not only help us to reduce our carbon footprint, but they actually help us save money.

"I keep bread which is past its freshness in the fridge and use it for toast!"

Sent in by:
Alma, West Oxfordshire

Tip 1

"I put out bacon rind and any fat from my frying pan into the garden to feed the birds."

Sent in by:
Louise, West Berkshire

19

"I plan my meals for the week before I shop. I always include two meals that are just from food that keeps in case of last-minute invitations. I save so much money (and time) and we waste less food."

"I use up Easter Egg chocolate by making it into fridge cake. This is also a good way to use up nuts and dried fruit left over from Christmas."

Fridge cake

250g of any biscuit that needs eating
300g any old chocolate that's lying around
100g unsalted butter
150g golden syrup
200g approx of whatever dried fruits or nuts need using up

Melt the choc, syrup & butter. Bash up the biccies and add them to the choc mixture. Stir in the rest of the ingredients, press into a loaf tin lined with cling film & leave it to set in the fridge

Tip 4

Gizmo the Girly

21

"I use up bread crusts. I make them into breadcrumbs for stuffing to go with my Sunday roast dinner or sometimes I dice them up and put them in the garden for the birds."

Tip 5

"I boil up my carrot tops and peel, leek trimmings and onion peel with my chicken carcass to make stock for soup or risotto."

Tip 6

"If I have bananas that are overripe I make banana bread or banana muffins. The bananas can be really black and mushy and the cakes will turn out just fine."

Sent in by: Pete Davis,
Founder of Part-Time Carnivore,
www.parttimecarnivore.org

Tip 7

Grave the Girl

"If I have flat lemonade I use it for cut flowers.   They like the sugar!"

Sent in by: Iris, Lancashire

"I eat the end of the cucumber. I've never understood why people waste half an inch off the end. It tastes fine to me."

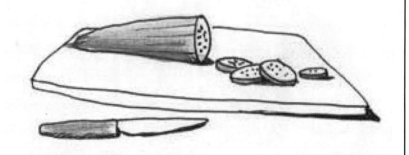

Tip 9

"When I buy things like sundried tomatoes or olives that come in oil I use up any oil left in the jars for stir frying or browning meat. It is good for added flavour. Then I rinse the empty jar before recycling it."

Tip 10

"I take in any leftovers for my lunch at work instead of buying shop bought, packaged sandwiches. I feel it is healthier, saves money and gives me more variety."

"When I buy pre-packed paté I put half the pack in the freezer until I've used up the first half, so that it doesn't go off before I can use it all up."

Tip 12

"I share vegetables like cauliflower and cabbage with my grandma because she wouldn't use up a whole cauliflower or even one of those small cabbages, but I can use up what she doesn't want."

"Ask your neighbours with lovely big fruit trees and too much fruit, to share it with yourself and the local community! No fruit goes to waste and you get to eat lovely local, low carbon fruit. And climb trees. Check out www.abundanceedinburgh.com for inspiration!"

Sent in by:
Valla Moodie, Edinburgh,
www.abundanceedinburgh.com

Tip 14

"I organise volunteers to collect surplus fruit from trees and bushes in gardens and on public land in our village, which we use at community events and workshops. It's a great way to improve sustainability, bring everyone together and also to raise money for other local events and activities."

Sent in by: Andrew,
www.facebook.com/
kidlingtonabundanceproject/

Saving Water

The Science

Even in a country like the United Kingdom where it rains a lot, water is still a precious resource.

Although 70% of the Earth's surface is covered with water, only about 1% of it is freshwater which we can turn into drinking water, water for washing ourselves, our environment, and for growing and preparing food.

There are many countries in the world where water shortages are a constant problem. There are places where some people get by on as little as 10 litres of water per day.

The average UK household uses around 500 litres of water every day and about a third of that water is flushed straight down the toilet! The carbon footprint of household water usage in the UK amounts to around 35 million tonnes of CO_2e per year.[4] This includes energy for heating water, energy required for abstracting, treating and supplying water, and for treatment of waste water. But all that is a drop in the ocean compared to our total water footprint.

Much of the water usage in the world – in fact around 70% of it – is for agriculture. A further 20% is used by industry and the remaining 10% is domestic usage.

Research has been carried out to identify the water footprint of avoidable waste food. The figure is 6.2 million cubic metres per year, which works out at 243 litres per person per day.[5] Remember that's just the stuff we needlessly waste!

It might surprise you that a pound of beef takes 6810

---

4 http://www.environment-agency.gov.uk/business/topics/water/109835.aspx

5 http://www.wrap.org.uk/content/water-and-carbon-footprint-household-food-and-drink-waste-uk-1

litres of water to produce – that's in the water the cow has drunk and the water required to grow the food that it has eaten. A pound of pork takes less than a third of that, requiring 2182 litres.

As you can see from the table below,[6] a pound of rice needs almost as much water as a chicken and it is clear why goat is a popular meat in dry countries.

Food (per lb)	Embedded Water Footprint (litres)
Beef	6810
Pork	2182
Chicken	1773
Rice	1700
Wheat	500
Goat	480
Potatoes	450
Corn	409

This shows us that one way of saving water is to cut down on the food we waste.

Of course it is not just our food that has a water footprint. What we wear, the energy we use in our homes, how we get around. These things all have a water footprint. That's water we don't control directly though, but the water we use in our homes *is* under our control and that's where we can have most influence.

6 National Geographic, environment.nationalgeographic.com/environment/freshwater/embedded-water

The Maths

Did you know, a cubic metre is equivalent to 1000 litres? Let's take a look!

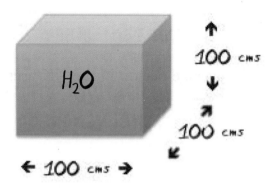

1 cubic metre = 100 x 100 x100 = 1,000,000 cm³

1 cm³ = 1 millilitre (ml) so

1,000,000 cm³ = 1,000,000 ml

There are a thousand millilitres in a litre so we divide it by 1000 to convert to litres.

1,000,000 ÷ 1000 = 1000

Why not take a look at one of your water bills? If your house has a water meter then you should be able to see how many cubic metres (m³) you have used over how many days.

For example, my household used 81 m³ in 180 days. Let's convert those cubic metres to litres:

81 x 1000 = 81,000

Now we can work out how many litres per day we used.

81000 litres ÷180 days = 450 litres per day

There are 5 of us in my house, so that means we each use 90 litres per day.

The Department of Health in the UK suggests that we need about 2.5 litres of fluid every day. However, some of that fluid comes from the food we eat and some is recovered by our bodies. So we need to drink around 1.5 litres of fluids. What do we do with the remaining 88.5 litres?

By the way, the average person in the UK uses 150 litres of water per day[7], so we are using only 60% of the average water use in our house. I'm puzzled, because we all drink water (or use water in drinks), we all wash and flush the toilet, we wash our clothes in a washing machine and we have a dishwasher. We use water for cooking and cleaning. What do people use the other 60 litres a day for?

But back to my own 90 litres of daily water usage. I have looked at roughly how fast the water flows through my taps. For each tap, I used the stop watch facility on my mobile to time one minute and put a bucket under the tap to catch the water. I then got another bowl and a litre measuring jug and carefully measured the water as I transferred it into the bowl so I didn't waste it. Then I used the water to clean the bathroom!

I've also thought about how often and for how long I do various things involving water. I have added a bit more onto my outdoor use of water. Although I occasionally use clean water for washing things outside – even sometimes my car – I also have several water butts from which, this year at least, I have had a constant supply of water for my garden. Not everyone has this facility to store rainwater. I put my information into an Excel spreadsheet and used the chart function to create a pie chart.

7 Department for Environment, Food and Rural Affairs

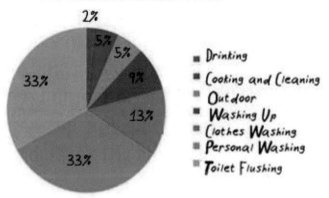

My average daily water usage

2%

5%
5%
9%
33%
13%
33%

- Drinking
- Cooking and Cleaning
- Outdoor
- Washing Up
- Clothes Washing
- Personal Washing
- Toilet Flushing

The pie chart shows me clearly where I use most water and where I could perhaps look at using less.

The average UK household water usage is 500 litres per day.[8] How come, if the average person uses 150 litres? Well, the calculation is based on a 4 person household but with a reduction for shared usage for things like cleaning, cooking, washing clothes etc. Let's take a look at what happens if every household in the UK reduces the amount of water they use by 10% per day:

10% of 500 litres = 50 litres

50 litres x 26,300,000 households = 1,315,000,000 litres

That's 1.3 billion litres of water saved every day!

Whilst researching these facts and figures I came across plenty of water saving advice, lots of which advocates water saving appliances in the home and that has a cost.

8 Environment Agency: http://www.environment-agency.gov.uk/homeandleisure/beinggreen/117266.aspx

But there were equally as many that would cost nothing at all. Simple advice like not flushing away tissues or cotton wool pads down the toilet, saves about 7 litres of water each time (the average water usage when you flush a toilet).

Consider this: if there are four people in a family and they all shower for four minutes instead of eight minutes[9], let's say four days out of seven when they are not washing their hair, how much water would that save? We'll use my calculation of water flow rate from earlier.

8 minutes x 4.5 litres per minute = 36 litres

36 litres per shower x 4 people x 4 days = 576 litres

whereas:

4 minutes x 4.5 litres per minute = 18 litres

18 litres x 4 people x 4 days = 288 litres

This gives a saving of 288 litres over the week. That's more than 40 litres per household off the daily allowance.

If you follow Carolyn's tip (Tip 16) and don't run the tap while you are cleaning your teeth, except when you are actually rinsing your brush and the sink, you can easily save the remaining 10 litres to get your household usage down by 10%.

Recently toilets have been designed to use less water per flush. Some time ago, water companies gave out free water hippos for people to put in their older toilet cisterns. If you have one you could perhaps try it out.

There is a national campaign to reduce the UK average water consumption by 20 litres per person per day. I think

---

9 Eight minutes is the UK average time spent in the shower according to the Unilever UK Sustainable Shower study.

we can do better than that, don't you?

A Word About Bottled Water

Mike Berners-Lee estimates the carbon footprint of a pint of tap water at 0.14g. He puts the figure for a 500ml bottle of mineral water at an average of 160g. That's more than 1000 times the footprint of tap water![10]

There are 195 days in a typical school year, but let's give an allowance for a few days missed and use the figure of 180 days. If there are two children in your family who take a refillable bottle of tap water to school every day, instead of buying a bottle of water or other drink in the canteen at a cost of 50p, you will save yourself a whopping:

$$2 \times £0.50 \times 180 \text{ days} = £180$$

It also gives a carbon saving of more than 57.5 kg which amounts to more than 650 hours of watching television or 4,791 hours of me using my laptop - and that's about how long it has taken me to produce this book!

---

10 Berners-Lee, M. (2010) *How Bad Are Bananas? The Carbon Footprint of Everything,* Profile Books

"When I'm brushing my teeth, I just use the tap water at the beginning and end, first to moisten the brush and at the end to rinse the brush and my mouth. I turn the water off when I'm actually doing the brushing because running the tap water in between is a waste."

Sent in by: Carolyn Miller,
Writer from Los Angeles and Santa Fe,
www.carolynmiller.com

"I have an old 2-litre squash bottle and an old 4-litre plastic milk carton by the sink, each with the top cut off to make a wider opening: when running the hot tap, I catch all the otherwise wasted cold water in these while I'm waiting for the hot to arrive. I then use it for adding a splash of cold to the bowl, watering my plants, rinsing out cans/bottles before recycling them, rinsing the sink, etc."

Sent in by: Rachel the Gardener,
www.rachel-the-gardener.co.uk

"I keep a jug of water in my fridge so I don't waste water waiting for the cold to come through."

Tip 18

"I plant my tomatoes directly into the ground rather than in grow bags. That way they need less watering because their roots can roam further."

Sent in by: Frank, Flackwell Heath

Tip 19

"My children take water to school in plastic bottles. They don't always drink it all, but instead of throwing the water out I now use it to water my plants. It's great! I'm no longer a house plant disaster."

Tip 20

"In dry weather I use a washing up bowl and, when I've finished washing up, I tip out the water onto my flower beds."

Tip 21

"I always turn off taps that are left running or lights that are left on. It is surprising how many people don't do that."

WHO LEFT THE TAP RUNNING??

Tip 22

"I use the water from boiling eggs to water my plants. The water contains valuable nutrients from the eggs that the plants appreciate."

Sent in by:
Valerie, France

"I take a jug into the shower to catch some of the hot water that misses me. I then use this hot water to shave in. Saves gas and water in one simple way."

Sent in by: Andy Redfern, Co-founder of www.ethicalsuperstore.com

Saving Energy

The Science

Most of the energy produced in the UK comes from resources that are considered to be finite, such as oil and coal. Although oil and coal are still being made through geological processes, the rate at which they are being made is very slow. On the other hand our demand for electricity has been increasing steadily. This means oil and coal are considered finite because it is predicted that we are likely to run out.

There are lots of sources of energy that are renewable – that is, they will go on being produced at a rate that means we could never use them up.

The sun is one such source. There is also: wind power; the power of moving water; the power of biomass, like wood or oilseed rape (as long as it is replanted); and biogas, such as that produced by anaerobic digestion of food waste and manure.

So if there's plenty of renewable energy available, what's the problem? Why all the talk about the need to conserve energy?

The sources of renewable energy are there but the ability to harness that energy is still limited. This is simply because we were building the infrastructure around the wrong sources of electricity, i.e. the finite sources of fossil fuels rather than the infinite sources like the sun, wind and moving water. More recently there has been investment in renewable sources of energy, but at the moment we still produce less than 4% of our energy [11] in the UK from renewables.

By 2020, the UK has a target of 15% renewable energy. This target can be achieved in 2 ways:

- *By reducing the overall demand for energy.*

- *By increasing the amount of renewable energy harnessed.*

The current carbon footprint of grid electricity is just over 500g per kilowatt-hour (kWh).[12] This includes an adjustment for the present contribution from renewable sources. Renewable energy, when generated on site (i.e. where it is used) is considered to have a zero carbon footprint rating. These carbon footprint ratings don't take into account the 'embodied carbon emissions' from the manufacture of the technology. For instance, in the case of solar panels it takes into account the installed solar panels but not their manufacture and installation. In the case of electricity from a coal-fired power station it doesn't take into account the construction of the power station itself.

Saving energy is one of the most cost-effective ways of reducing carbon emissions. It is estimated that up to 40% of the energy we consume is actually wasted, so just by making some common sense changes to our lifestyles, without any reduction in quality, we could save a considerable amount of energy. Simple everyday actions

---

11 The Department of Energy and Climate Change (DECC) National Renewables Statistics puts the 2011 renewables contribution to total UK energy consumption at 3.8%.

12 The Defra/DECC Energy Conversion Factor for grid electricity as of August 2011 is 0.5246 kgCO$_2$e.

like not leaving lights on when we don't need them, not leaving phone chargers plugged in all night, waiting until the dishwasher or washing machine is full before turning on, drying clothes outside on a washing line instead of using a tumble dryer, all add up to a considerable carbon saving and save money too.

The Maths

Let's take a look at the amount of money you could actually save in one year.

A plugged in mobile phone charger continues to draw electricity even when your phone is not attached to the end of it or is fully charged. This is clearly a waste of electricity, but the amount is tiny.

It seems that a mobile phone charger may use 0.009 kWh per day. So if it was left plugged in for a whole year that would amount to 3.285 kWh which at 13p per unit amounts to around 43p.

$$0.009 \times 365 \times 13 = 42.705$$

However, a tumble drier which generally uses 3 kilowatts per hour (3kWh) costs 39p every time you use it for an hour. My tumble drier seems to dry my washing in about an hour and 20 minutes, so actually it costs me about 50p each time I use it. Let's say your family does 3 lots of washing a week that you manage to dry outside on the clothes line instead of in the tumble drier. That will save you:

$$£0.50 \times 3 \times 52 \text{ weeks} = £78 \text{ over the year.}$$

I hate ironing soooo much!

Add the fact that clothes that are line dried seem to need hardly any ironing, I think that's definitely worth the saving!

I've chosen these two different calculations not to belittle the importance of not leaving phone chargers plugged in all the time, but to make sure that we don't all start thinking we can save the planet by unplugging our phone chargers if our washing is dried by a tumble drier three times a week. Instead we all need to be using that free resource of wind power and mucking in with the chore of hanging out the family undies on the line - and bringing them in again when they are dry!

But just because maths is always fun when the numbers get big, let's look at the cost of every household in Britain leaving two phone chargers plugged in all the time.

£0.43 x 2 chargers x 26,300,000 households = £22,618,000

Hmm! That's enough to make me unplug mine.

"Switch all electrical equipment off, don't leave it on standby. Electrical equipment draws a lot of power if the LED is on/ flashing. I discovered this when I left my fully charged laptop on standby, and three days later it was on half power!"

Sent in by: Leo Thompson,
Teacher from Vienna, Austria

Tip 25

"Why use an expensive tumble drier when you can use the power of nature? Wind and sunshine is the cheapest and most environmentally friendly way to dry washing, so I hang mine outside."

Sent in by: Bev Hockley,
Customer Care Team,
www.goodenergy.co.uk

"Use a measuring jug to fill your kettle. Measure out how much water the things you use most often need (e.g. your teapots, mugs, flask, hot water bottle, etc) and then use the correct amount every time you boil the kettle. Even my 19 and 21 year olds have finally learnt to do this! (Still working on the husband...)"

Sent in by: Elizabeth Long, Housewife and mother of 4 from Kingston Lisle

Tip 27

"I charge up my mobile phone in the car. That way it is totally free as the electricity is a by product of the car's movement."

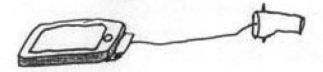

Tip 28

"I always try to fold my washing straight away when I bring it in from the line. That way I hardly ever need to do any ironing."

Tip 29

"I often hang clothes on the line on a windy day without washing them. Airing clothes rather than washing will help them last longer as washing damages the fibres."

Tip 30

"If you cook with electricity, you can turn your hob off before you have finished cooking so you use up the residual heat. It took me a while to get into the habit when I changed from gas but I've got used to it now!"

Sent in by: Sarah Ellison,
Twitter: @RecommendedOXON

Tip 31

"I do my week's ironing in one go rather than in bits because of the energy it takes to heat up the iron. I listen to an audio book from the library while I'm doing it."

Tip 32

"I close my curtains at dusk to keep the heat in. I also keep doors shut to rooms I don't need to heat all the time. That way I keep the rest of my house cosy without running up my gas bill."

Sent in by: Eileen from Peterborough

Tip 33

"When using food from my freezer I try to remember to take it out the night before rather than having to defrost it in the microwave."

"I use my laptop plugged in during daylight hours and then use up the battery in the evening because I have solar panels. That way I'm making maximum use of my renewable energy and maximum savings too."

Tip 35

"When I make a casserole that takes a lot of cooking, I sometimes make more than I need and freeze a portion."

Tip 36

"I managed to save on heating bills by blocking off draughts under my doors. You don't have to buy fancy draught excluders. Mine range from an old towel, some unused cushions, to a row of my children's abandoned stuffed toys!"

Tip 37

Saving Fuel

The Science

We understand now that fossil fuels are finite resources. But we rely more and more on being able to travel around the country and around the world. The global economy we have today relies on buying and selling goods all over the world. This means we use a lot of fuel.

It is likely that in the future we will use different kinds of fuel to move our goods and ourselves around the world.

Already we know that it is possible to run cars on the used oil from fish and chip shops. But there is not enough used chip shop oil to run all our vehicles. So it is not a viable solution, even though it does provide a good way of using a waste product and a far better way of dealing with the used oil than pouring it down drains.

Waste oil from chip shops and other food retailers is made into a form of biodiesel by straining it to remove the food particles and adding alcohol to make it combustible.

Biodiesel is also made from crops grown specifically for the purpose, such as rapeseed, mustard seed or soybeans. Today many oil companies sell diesel that is mixed with around 5% of biodiesel. However, trying to replace all the world's diesel requirements with biodiesel from crops would require huge amounts of land. Much of that same land is required for growing food crops, and deforestation is already a huge concern.

In his book *How Bad Are Bananas? The Carbon Footprint of Everything*, Mike Berners-Lee calculates that one hectare of deforestation has the carbon equivalence of 500 tonnes of CO_2e. In other words it creates the equivalent emissions to an average car driving 28 times around the world, or more than 108 return flights from London to Hong Kong. There are concerns that biodiesel from crops could increase greenhouse gas emissions, compared to fossil fuels, if not carefully managed in terms of land use and processing. We can't grow crops for biodiesel on land that we need for food production and we can't clear more land either. We also have to think about the crop requirements. Some crops in certain places need water and fertiliser to such an extent that the energy gained from the biodiesel is actually less than the energy required to produce the crop.

Researchers are looking into alternative ways of producing biodiesel. For example, technologies are being developed so that more of the plant can be used to produce the diesel, making the production less wasteful and providing more fuel per hectare of land. They are also finding crops that don't deplete the land of nutrients and so require less fertiliser, and they are looking at using the waste parts of food crops. In future, biodiesel production from algae may become cost-effective too.

So although biodiesel can replace a small amount of petrochemical diesel, it is not the entire solution.

So what other sources of fuel can we use?

Liquid Petroleum Gas (LPG) or Propane

LPG, such as propane or butane, is a mix of hydrocarbon gases, produced by refining petroleum or natural gas. It is thought to burn cleaner than petrol or diesel, producing fewer particulates. It produces around 14% less carbon dioxide[13] because there is a lower carbon to hydrogen ratio

---

13 http://www.direct.gov.uk/en/Environmentandgreenerliving/Greenertravel/
Greenercarsanddriving/DG_191580

compared to petrol. It also outputs less sulfur and fewer metal pollutants. However, this is still a fossil fuel, of which there are finite resources.

Electric Vehicles

There has been an increase in popularity of electric vehicles, although these vehicles still have a limited range and electric charging stations are not yet available throughout the UK.

Although electric vehicles do not produce emissions themselves, we should remember that much of our electricity still comes from fossil fuels and emissions come from the power stations where the electricity is made. As more clean electricity is generated, and technology improves, electric vehicles may become a suitable alternative.[14]

Hydrogen

Running cars on hydrogen produces no carbon emissions. But although hydrogen is one of the most common elements on earth, it is never naturally in its pure form. Hydrogen has to be extracted for example from water, using a process that requires energy. So until more of our energy comes from low carbon sustainable sources, hydrogen powered vehicles will not be more environmentally friendly than petrol or diesel vehicles.

Until sustainable sources of alternative fuels are found, we need to limit the rate at which we use up reserves of fossil fuels and that means we need to try not to waste fuel.

---

14 CAT, *Zero Carbon Britain 2030*

We need to think about our necessary travel and try to combine journeys. For instance, we can give them more than one purpose, such as by doing the shopping trip in the same journey as the school run. We can share with other people who need to make similar journeys, e.g. to school, work, sporting activities etc. We can use the most efficient means of transport and we can try to avoid unnecessary miles. All this will help our fuel stocks to last longer and mean that the alternative fuels are providing a higher percentage of our requirements.

The Maths

Transport is responsible for 24% of UK domestic emissions, producing around 130 million tonnes of CO_2e per year.[15]

We can reduce these emissions by transferring to more carbon-friendly modes of transport e.g. exchanging a long-distance car journey or a domestic flight for a train journey, or swapping our 3-4 mile car journey to a carbon free bike ride.

Car-pooling and lift sharing are perhaps the easiest way to save carbon, save money and save time all in one go. Sometimes lift sharing adds miles to a journey and adds time, but it will work out to be more efficient overall.

You could follow Tip 42 and get together with a few families to split up the transport to various sporting activities.

Here's an example:

Family A from Hopton: 2 children play football in Bigton on Tuesdays and Thursdays, 1 of whom plays hockey in Bigton on Wednesdays.

Family B from Skipton: 1 child plays football in Bigton on Tuesdays and Thursdays.

Family C from Skipton: 1 child plays hockey in Bigton on Wednesdays.

Skipton and Hopton are 2 miles apart. They are both 5 miles from Bigton. It takes 8 minutes to drive from Skipton to Hopton or vice versa. It takes 20 minutes to

15 CAT, Zero Carbon Britain 2030

drive from either Skipton or Hopton to Bigton.

Family A's journeys would be as follows:

	Miles	Time
Tuesday	10	40
Wednesday	10	40
Thursday	10	40
Total	30	120

Family B's journeys would be as follows:

	Miles	Time
Tuesday	10	40
Wednesday	0	0
Thursday	10	40
Total	20	80

Family C's journeys would be as follows:

	Miles	Time
Tuesday	0	0
Wednesday	10	40
Thursday	0	0
Total	10	40

If all three families get together and lift share, their week could look like this:

Family A's journeys over 2 weeks would be as follows:

	Miles		Time		Average/Week	
	Week A	Week B	Week A	Week B	Miles	Time
Tuesday	14	14	56	56	14	56
Wednesday	14	0	56	0	7	28
Thursday	0	0	0	0	0	0
Total	28	14	112	56	21	84

Family B's journeys over 2 weeks would be as follows:

	Miles		Time		Average/Week	
	Week A	Week B	Week A	Week B	Miles	Time
Tuesday	0	0	0	0	0	0
Wednesday	0	0	0	0	0	0
Thursday	14	14	56	56	14	56
Total	14	14	56	56	14	56

Family C's journeys over 2 weeks would be as follows:

	Miles		Time		Average/Week	
	Week A	Week B	Week A	Week B	Miles	Time
Tuesday	0	0	0	0	0	0
Wednesday	0	14	0	56	7	28
Thursday	0	0	0	0	0	0
Total	0	14	0	56	7	28

Family A's average weekly mileage has gone down from 30 to 21, and the time the parents spend driving to and from activities reduces from 2 hours a week to an hour and 24 minutes.

They save 9 miles a week and 36 minutes each week.

Family B's average weekly mileage has gone down from 20 to 14, and the time the parents spend driving to and from activities reduces from 1 hour and 20 minutes a week to 56 minutes.

They save 6 miles and 24 minutes each week.

Family C would have driven 10 miles a week. But by lift sharing, the parents now only have to go out every other week, freeing up almost an hour of their time. Their average weekly mileage has come down from 10 miles to 7 miles, which is a 30% reduction.

"Instead of jetting about the world to conferences, why not participate in them via social media. I went to the Virtual Frankfurt Bookfair recently: followed the debate via various websites, caught the gossip on Twitter and enjoyed posting imaginary news of the people I'd met and the parties I'd virtually attended. Just think of the saving in expenditure on travel costs and hot air!"

Sent in by: Chris Meade,
Director of if:book UK,
www.futureofthebook.org.uk

"I use a bike to move around (well, if it's not raining). It's eko, healthy and fun!"

Sent in by: Jaka from Ljubljana. I ride a bike and sometimes I write poetry. www.jaka.org

"I buy local food – it's fresher, lower in carbon and supports the local economy! Local non-organic is better than organic that has travelled miles."

Sent in by: Simon Kenton, from the Community Action Groups project in Oxfordshire, www.cagoxfordshire.co.uk

"I always remove my roof box and roof bars after using them for holidays because they cause additional drag and that leads to increased fuel consumption."

Tip 41

"I lift share with five other families to take our children to various sporting activities. That way I've halved the number of miles I have to drive each week."

Tip 42

"I live a long way from the shops and don't have a milkman, so to save making a special journey for things like bread and milk I buy my week's worth and freeze it."

Tip 43

"I live a mile and a half from my children's school, but we always walk when we can. We take the dog too. On the way we do spellings and times tables and have a good chat. Exercise, dog walk and homework all sorted by 9.30 a.m."

Tip 44

"I make sure I empty out my car boot. Extra weight in your car means extra fuel."

Sent in by: Chris,
www.veggiepower.co.uk

Recycle it

The Science

More and more stuff is being recycled in the UK, but the 2010/11 figure was still only 41.5%. This was an increase of 1.8% from the previous year, but we still need to improve on this.

There are still many people who feel that recycling is not important. But we can't keep making more and more stuff out of finite resources. It makes sense to use materials we already have rather than always making things out of new materials that we extract from the earth. Making things out of recycled materials takes less energy than making them out of new materials. It also means we need less space in which to dispose of our rubbish if we have less of it due to increased recycling.

In the UK, many people have some form of recycling collection from their home, organised by their local council. The materials accepted for recycling vary across the country depending on the availability of local technology for collecting, separating the materials and recycling them.

There are two different kinds of collection: 'kerbside sort' or 'co-mingled' collections.

The kerbside sort uses a recycling lorry with various different compartments for the different materials. Households have small boxes or bags in which to separate out their recycling and the collection crew further separate the materials into the relevant compartment of the lorry. Materials are returned to the recycling yard and further sorted by machinery, which separates, for example, the steel cans from the aluminium cans and plastic containers.

Co-mingled collections are where households are provided

with a large bin, usually a wheelie bin, in which to put all their recyclable materials together. These are emptied by the collection crew into the recycling lorry and taken to a 'Materials Recovery Facility' (MRF). Here, they are sorted into different materials by passing along conveyor belts and through various machines designed to separate out the different types of packaging, diverting it towards the appropriate baling machine. There are still people involved in this sorting process, as it needs human beings to pick out the things that can't be recycled.

As technology improves and places are found for the sorted materials to be sold to, more and more materials are likely to be collected. It is a good idea to read any new information that is sent out by your local council or your recycling team as it may tell you about new materials that can be collected, that previously couldn't be recycled in your area.

The most important element of recycling and of waste management in general is for people to keep their food waste separate from their other waste. When food waste is kept away from packaging materials, the materials stay clean and maintain their quality. Food waste rots and contaminates the recyclable materials. Although there are processes that can remove the contaminants, the whole recycling operation will be cleaner and more cost-effective when food waste is kept separate. Besides, as you may remember from the earlier section on reducing food waste, there are better ways of dealing with it. When food waste is kept separate, it can be composted or sent for anaerobic digestion and this brings value from it.

Making space for keeping recyclables, food waste and non-recyclable waste separate in the home makes it easier to recycle more. Many households in the UK have been provided with 'food waste bins'. They often consist of a smaller food caddy to be kept in the kitchen and a larger vermin-proof bin that is kept outside. Some people line the food caddy with compostable bags, but you don't have to do this. You can line it with a couple of sheets of newspaper or you can put the food waste directly into the bin. It is important not to use plastic shopping bags as a liner because these do not break down in the same way as your food waste and they will result in unnecessary contaminants in the food waste recycling system.

In addition to your food waste bin it is easiest if you have two further bins: one for recyclable waste and one for non-recyclable waste. You need to wash tins, plastic tubs and plastic packaging to remove any food left in them. Rinsing drinks bottles and cans helps but it is not essential. It is more important that every bottle and can gets into the recycling, than only recycling some of them because you feel you don't always have time to rinse them all.

You do need to empty the remains of your shepherd's pie from the plastic or foil tray into your food waste bin before you recycle the tray! It may surprise you, but there are lots of people who don't bother. You can't throw the bag from your favourite fast food outlet straight into the recycling bin, complete with picked out bits of gherkin and tomato ketchup-covered chip carton. But most packaging, if treated in the right way, is recyclable. So, you see, it is down to us to do our bit.

What happens to the materials we recycle and what difference to the environment does it actually make?

Ideally, recycled materials would be made back into the same product and in doing so the recycling process saves on raw materials and saves energy. But this is not the case with all products. Sometimes recycled packaging is turned into different products. Sometimes this is called up-cycling and sometimes it is down-cycling. The 'up' or 'down' bit refers to whether the product has a higher or a lower value than the packaging material it was made from. This has the advantage of making use of otherwise waste resources, and removes the need to send it to landfill. An example of down-cycling is using recycled paper to make toilet paper. It is down cycling because it has a lower value than paper for printing and because it can only be used once in its down-cycled state. An example of up-cycling is turning plastic drinks bottles into the filling for ski jackets. Again, up-cycling removes the need to send the plastic bottles to landfill. The problem with up-cycling is that there is less demand for the up-cycled product than there is for the packaging product. Taking the example of plastic bottles and ski-jackets: it takes five 2-litre plastic bottles to make enough 'fiberfill' for one ski-jacket. I think my rate of consumption of products in plastic bottles is on average 5 per month, whereas I am only likely to need 5 ski-jackets in my whole lifetime if you include the ones I grew out of and passed on to others as a child.

So the best recycling for packaging is where it can be turned back into packaging, thus saving raw materials as well as energy.

Aluminium is the most easily recycled packaging material. It can be recycled over and over again, and requires very little energy to make it reusable as more aluminium packaging. It is simply melted and crushed and added into the fresh aluminium. Recycling an aluminium can takes less than 5% of the energy that it takes to make an aluminium can out of the raw materials.

Glass bottles and jars can also be recycled over and over again into new bottles and jars. There are many other uses for recycled glass too. For example, glass cullet - that's the crushed glass from the recycled bottles and jars - can be used in building materials like concrete and in kitchen work surfaces. It takes around 30% less energy to produce a glass bottle from recycled glass than it takes to produce one from the raw materials. However, better than recycling glass is to reuse it. Glass bottles and jars can easily be sterilised for reuse with minimal energy usage. Most reuse systems involve picking up empties when the filled bottles are delivered, making for efficient transport as well.

Paper can be recycled about 6 times before the fibres become too short to make back into paper. Recycling paper takes around 45% less energy, and around half the water than making paper from new wood pulp. Recycling one tonne of paper is estimated to save on average 17 trees. Recycling cardboard halves sulfur dioxide emissions and uses 25% less energy compared to making cardboard from new wood pulp. Not only that, for every tonne of cardboard recycled you save 6.8 cubic metres of landfill space.

Plastic packaging comes in numerous types and qualities.

The most easily recycled are Polyethylene Terephthalate (PET 1), for example the 2 litre clear plastic bottles that you buy fizzy drinks in, and High-Density Polyethylene (HDPE) like the white plastic milk bottles. Polystyrene (PS6), which is used to make things like yoghurt pots (not the polystyrene used for plastic cups and some meat trays which is actually extruded polystyrene), is now being more widely recycled too. Every tonne of recycled PET (rPET) produced saves over 1.5 tonnes of CO_2e.

Lots more things can be recycled too, so take a look at the tips and see what you could be doing to turn your rubbish into someone else's resources.

The Maths

What's in your recycling box this week?

Here is an example of how much energy you can save by recycling your household waste. The table below uses a tool provided by Coca-Cola on their website. They call it their 'Recyclometer'. It takes the contents of a recycling box and shows the energy saving from recycling this stuff in terms of the energy required to power a television.

A recycling box contains	Minutes of television
1 cereal box	90
2 aluminium cans (330ml)	210
1 egg box	31
1 glass wine bottle	90
1 magazine	360
5 items of junk mail	90
1 newspaper	240
2 500ml plastic drink bottles	54
2 2-litre plastic drinks bottles	90
2 steel food cans	120
1 cardboard toilet roll tube	6
Total minutes of TV	1381

The energy saved from this recycling is enough to watch a television for around 23 hours[16].

Why don't you see what your recycling could save? Go to http://www.coca-cola.co.uk/environment/recyclometer.html and give it a try.

---

16 Calculated using the Coca Cola Recyclometer. http://www.coca-cola.co.uk/environment/recyclometer.html

Let's do another calculation.

Recycling one tonne of plastic drinks bottles saves 1.5 tonnes of carbon.

If a school has 1500 students and on average one third of them buy a drinks bottle each day, which they then recycle, they will save:

1500 students ÷ 3 = 500 bottles purchased each day

There are 195 days in a school year. So let's work out how many bottles will have been purchased by the end of the year.

500 x 195 = 97,500

One 500ml plastic drinks bottle weighs 24g[17], so that means a total weight of 2,340,000 grams (or 2.34 tonnes). Let's work out the carbon saving by multiplying it by 1.5. That gives us a carbon saving of 3.51 tonnes.

We could work it out in terms of TV hours. To do this you take the number of bottles, multiply it by the number of minutes the recyclometer tells us you can get for each bottle, which is 27 minutes.

97500 x 27 = 2,632,500

To see how many hours that is, we have to divide this by the number of minutes in an hour, i.e. 60.

2,632,500 ÷ 60 = 43,875 hours

How many days is that? Let's see.

43875 ÷ 24 = 1828.125 days

Wow! that's just over 5 years worth of watching TV.

---

17 Weight of a 500ml Coco-cola bottle without lid.

"I keep an old toy box on my landing to collect up recyclables from upstairs, so I don't have to sort through my bathroom & bedroom bins."

Tip 46

"I rinse out my tin cans then crush them flat by standing on them, so they take up less room in my recycling box."

Some cans you take the bottom off with a tin opener. Turn the can on its side on the floor, place the top and bottom inside, then it can easily be crushed flat keeping the sharp edged bits safely inside.

Tip 47

Gromit the dork

"When I upgraded my laptop I took my old one into our local repair shop where they repair parts and reuse them or recycle them."

"When I occasionally have large cardboard boxes I cut them up with strong kitchen scissors so they fit in my recycling boxes, which saves me driving to the tip with them."

Tip 49

"I wash my aluminium foil and reuse it. Sometimes I put it through the dishwasher. That way I can reuse it three or four times before I then scrunch it up to put it in my recycling."

Tip 50

"When I'm out and about I always take any packaging home with me so I can reuse it, recycle it or compost it."

Sent in by: Ben Molyneux,
Founder of The Oxfordshire Project,
www.oxfordshireproject.co.uk

Tip 51

"I tear my name and address and any confidential information off my post and put the torn up scraps into my composter or into my fire. I recycle the rest of the paper."

Tip 52

"I crush my plastic bottles and then put the top back on. This means they take up less room in my recycling bin and it saves space in the collection lorries too. That saves miles, saves fuel and saves time."

Note: It is important to crush the bottles to get the air out before you replace the cap.

"**Whenever I get off the train, I scoop up all the free papers that passengers have left on their seats (or on the floor!) and take them home with me to make sure they get recycled.**"

Sent in by: Ian Skillicorn,
Founder and Producer of
www.shortstoryradio.com

Tip 54

"I wash out and keep foil containers and have one by my sink to collect foil milk bottle tops, the aluminium foil from sweetie wrappers, the foil lids from tins of cocoa powder etc. I scrunch this all up together to make sure it gets recycled."

Note: It is important that what you scrunch together is all aluminium and not any other kind of material. You can scrunch it all up in a piece of tin foil too. Wash the foil first!

"Oxfam and maybe other charity shops recycle old mobile phones for people in Africa, so I donate mine to them."

Sent in by: Alma,
West Oxfordshire

"I collect beer bottle caps into a washed out baked bean tin to make sure they get into my recycling box. My local council collects bottle tops in the kerbside box, but if yours doesn't, there may be a local scrapstore that you could give them to."

Tip 57

"My local library recycles all used printer cartridges, regardless of type or make :-)"

Sent in by: Alma,
West Oxfordshire

"Old batteries are collected by the local council for recycling as are textiles which are beyond passing on :-)"

Sent in by: Alma,
West Oxfordshire

Tip 59

"I add water to my shampoo bottle when it gets near the end and it lasts way longer. Then, when I really have got every last bit out, I recycle the bottle."

Sent in by: Janet,
West Oxfordshire

Tip 60

"I collect used batteries into a small bag in one of my kitchen cupboards. They have to be kept separate in the recycling."

Note: If your council doesn't accept batteries in the recycling picked up from home, look out for battery recycling points in supermarkets. Batteries shouldn't go to landfill.

Tip 61

"I squash all my flyaway plastic into a bread bag and tie it up before I put it out into the recycling. Loose bits of this kind of plastic can get stuck to other materials. This keeps it out of the way."

Some local councils don't accept flyaway (or film) plastic. If your council is one that doesn't, check out your local supermarket. You can also try reducing the amount you acquire and reuse where possible.

Tip 62

Gizmo the Geek

Reuse
it

The Science

You may have heard of something called the Waste Hierarchy. The expression 'Reduce, Reuse, Recycle' is part of that hierarchy. The Waste Hierarchy is a framework for how we need to deal with waste. Since 2011 it is a legal requirement for businesses to apply this framework.

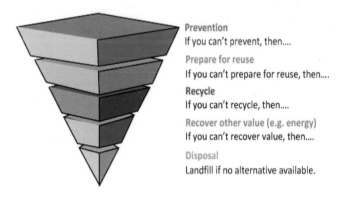

Prevention
If you can't prevent, then....

Prepare for reuse
If you can't prepare for reuse, then....

Recycle
If you can't recycle, then....

Recover other value (e.g. energy)
If you can't recover value, then....

Disposal
Landfill if no alternative available.

The Waste Hierarchy[18]

This hierarchy ranks the waste management options in order of environmental impact.

At the top of the scale is reducing or 'preventing' waste. The best thing we can do is to try to prevent waste. Once waste has been created, what we should be aiming for is to prepare the waste for reuse.

If the waste can't be reused then it should be recycled. Only if it can't be recycled should it be used for energy recovery, for example where household waste is incinerated to generate electricity or food waste is sent for anaerobic

---

18 Image reproduced with kind permission of Scottish Environmental Protection Agency: http://www.sepa.org.uk/waste/moving_towards_zero_waste/waste_hierarchy.aspx

digestion to produce gas and fertiliser.

Only as a last resort should disposal be considered, i.e. sending waste to landfill or incineration without energy recovery.

Reuse sits near the top of the scale above recycling because the environmental footprint is lower. The energy requirements are lower for reuse than for recycling, and consequently the carbon footprint will be lower.

For example, it takes less energy to collect bottles on a return journey from a delivery, and wash them and sterilise them for reuse, than to collect them on a separate journey, take them to a glass recycling facility, break them down into glass cullet and reform them into new glass bottles. Glass milk bottles are used an average of 13 times.

Like recycling, reuse saves resources and it reduces the amount of waste being sent to landfill.

Charity shops, jumble sales, swap shops, Freecycle or Freegle, eBay, and lots more provide ways to reuse many household items.

According to HarperCollins, libraries can lend books an average of 26 times, though librarians consider this to be vastly underestimated, and an average figure of 40 times has also been suggested[19]. I wonder how many times LoveFilm.com lends out a DVD?

---

19 Source: http://news.bbc.co.uk/1/hi/programmes/click_online/9421996.stm

The Maths

UK supermarkets have recently led highly successful campaigns to encourage their customers to switch to reusable shopping bags. Many supermarkets monitor bag reuse and award loyalty points each time a customer reuses a bag. This enables them to collect statistics on how many bags have been saved via reuse.

M&S used to give out over 460 million food carrier bags a year. The usage dropped by 81% following the introduction of their 5p per bag charge.[20]

81% of 460,000,000 = 372,600,000

That's a lot of carrier bags saved. Each bag has a carbon footprint of 10g[21].

372,600,000 x 10g CO_2e = 3,726,000,000g

There are 1,000,000 grams in a tonne.

3,726,000,000g / 1,000,000 = 3,726 tonnes

So 3726 tonnes of CO_2e have been saved by people reusing their plastic bags. And that's just at Marks and Spencer!

---

20 Source: http://plana.marksandspencer.com/we-are-doing/waste/stories/24/

21 Source: Berners-Lee, *How Bad Are Bananas? The Carbon Footprint of Everything*

Sainsbury's have given out 563 million loyalty points for bag reuse which suggests a carbon-saving of 5,630 tonnes.

If we add the two amounts together we get:

$$3726 + 5630 = 9356$$

We often see carbon savings shown as an equivalence to taking a certain number of cars off the road. We do this because we find it hard to imagine what a tonne of greenhouse gas amounts to, but we understand that cars use fuel and that they emit pollution. Because cars are an everyday part of our lives it is easy to imagine what 100 cars looks like and what 1000 cars looks like or what 10,000 cars looks like. And we can imagine what our roads might be like with 100 or 1000 or 10,000 fewer cars on them. Scientists have worked out that one car emits around 5.1 tonnes of CO_2e gases in a year.[22]

So that's how they can tell us that by reusing their plastic bags, customers at these two supermarkets have made a carbon saving equivalent to taking 1835 cars off the road for a year.[23]

22 U.S. Environmental Protection Agency: http://www.epa.gov/cleanenergy/energy-resources/refs.html#vehicles

23 The 5.1 tonnes of CO_2e is rounded up.

"Use your shredded paper in your rabbit hutch to keep your pets warm. You can feed them all your vegetable peelings and apple cores too."

Sent in by: Stephanie Hale,
Entrepreneur and Author,
www.millionairewomenmillionaireyou.
com

Tip 63

"Sort and save all your flat-pack extras (leftover screws, wooden plugs, fixing braids, brackets, spanners, etc) in old transparent Tupperware containers. You'll be amazed at how many little jobs they'll serve for later, freeing you from trekking down to the hardware store to buy new ones. This saves fuel, time and money, as well as reducing waste."

Sent in by: Tia,
London

"If I want to send anything via snail mail, I reuse envelopes sent to me - all shapes and sizes. I haven't bought a new envelope in donkey's years!"

Sent in by: Christine Wilks,
Digital Writer and Artist,
www.crissxross.net

Tip 65

"I use Ecomodo.com to lend and borrow tools and equipment that I only need once in a while. Things like a sledgehammer, wallpaper remover, lawnmower, wheelbarrow, sewing machine, tools and cookware are on offer in my area. Ecomodo saves me money, storage space and effort by allowing me to use large, expensive and specialised tools without having to own them."

Sent in by: Kate,
Scientist, Oxford,
www.ecomodo.com

Tip 66

"I keep my shopping bags in the boot of my car. My house is small so it means they are out of the way and I have them when I need them."

"I put empty margarine tubs and ice cream tubs in the dishwasher and then use them when my children take cakes or sausage rolls to parties. That way I don't need to worry about getting the box back."

here's the sausage rolls!

– Mmm, and there's the ketchup!

Tip 68

"When I stay in hotels I always bring back the soap with me because otherwise it would just get thrown away. When I unwrap it I keep the packaging in my wash-bag so I have something I can wrap it in to take it home."

Sent in by: Richard, from Oxfordshire

Tip 69

"I buy my books second hand from charity shops or borrow them for free from a library. It saves me a lot of money."

"I always carry a couple of the old-style carrier bags (because they fold up so small) and a 'bag for life' in my handbag. Each time I unpack the bag it goes straight back in my handbag."

Tip 71

"Rather than buying sandwich bags for my children's packed lunches I reuse the bags from apples, pears, and bread. I did wonder if they'd complain but I don't think they've even noticed."

Tip 72

"When I hang out my swimsuit after going to the gym I hang out the plastic bag too. I reuse the same bag over and over again."

Tip 73

Gizmo the Geek

"We use our own thermos type mugs at work for when we get coffee from coffee shops in town. We take them to conferences with us too. It saves using throw-away paper or plastic cups and saves coffee/tea because it stays hot much longer so we don't have to throw away the cold dregs."

soya latte, please.
- And that there's my take away cup!

Sent in by:
Richard and Marlin Armstrong,
Directors of Heavenly Spa in
London and Coaches at
www.publicspeakersuniversity.com

Tip 74

"I put nettles that I pull up from my garden into an old bucket. I leave it to fill with rainwater and rot down for a few weeks and it becomes a powerful fertiliser. It is free and there's no packaging. (But it stinks a bit so don't be tempted to stir it!)"

Sent in by: Chris,
www.veggiepower.co.uk

"I use my grass clippings as a mulch to suppress weeds in my flowerbeds. It helps to conserve the moisture in the soil too."

Tip 76

"I use up birthday cake candles that are too short to use on a cake to light my fire because I'm not very good with matches. It saves me loads of matches and saves me binning the candles."

Tip 77

"When I go out for coffee or to eat I usually bring home the napkin if it is barely used. I keep them by my compost bin at home and use them to wipe out greasy pans before I put them in the dishwasher."

Tip 78

"I use Freegle to have a good clear out now and then."

Tip 79

"You can turn bike inner tubes into very strong rubber bands with a pair of scissors. It's a good idea for when your tube has had one too many patches, however bike repair places throw loads of them out and you can get as many as you want for free."

Sent in by: Douglas,
born in Glasgow studying in Edinburgh,
www.abundanceedinburgh.com

Tip 80

"I use banana skins to fertilise my roses. They like the potassium."

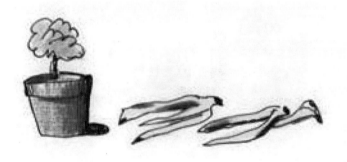

Sent in by: Valerie, France

Tip 81

"I keep any broken crockery in a bucket in my garage to use when potting up plants that I'm dividing or growing from seed."

Sent in by: Sonia, from South Wales

Tip 82

"I compost my teabags and coffee grounds."

Tip 83

"I turned an old water trough into a weed trap for my garden. Let it fill with rainwater and add a bit of pondweed if you know someone who has some. Then you can throw in all your troublesome weeds like docks and bindweed so you don't need to have bonfires."

Tip 84

"I empty my hoover into my compost bin."

Tip 85

"You sometimes find that clothes labels are attached with ribbons. Boxes of chocolate, perfume bottles and makeup sets sometimes have ribbon round them. I collect all these bits of ribbon into my ribbon box and so I've always got something to use when wrapping presents or giving bunches of flowers from my garden."

Tip 86

"I get together with friends occasionally for a clothes swap party. We all bring along clothes that no longer fit or we don't like anymore and people try them on. It is nice to see an old favourite being given a new lease of life and it is great to come home with a new outfit without spending any money."

Sent in by: Helen Lederer,
Actress, Writer and Comedian,
www.helenlederer.co.uk

"I try and use the tumble dryer as little as possible but after use, the lint from the dryer can be composted."

Sent in by: Alma, West Oxfordshire

Tip 88

"I add the ash from my wood burning stove to my compost heap. I also use a bit of cold ash pressed onto kitchen roll to clean the window of the stove every now and then."

Sent in by: Valerie, France

Tip 89

"When I buy fruit and vegetables like peppers, courgettes, carrots and leeks I don't bother with a bag. When I buy vegetables like sprouts or mushrooms I put the bags back into my shopping bags, so if I do need vegetable bags I usually have one I can reuse."

Tip 90

"I have a permanent 'charity shop bag' on the go to collect stuff I don't want anymore."

Sent in by: Jilly, North Yorkshire

"I cut my toothpaste tubes in half and rinse them out in the sink. Then I give the sink a good scrub round using the toothpaste. It gets my sink nice and clean and smelling fresh."

Tip 92

"I only buy clothes from charity shops. You get some amazing finds and it saves me so much money. But I probably do buy too many clothes, so sometimes I pass them on to friends or take them back to the charity shop when I've got bored with them."

Note: The buying bit does, of course, involve parting with your money, but the giving bit is totally free!

Tip 93

"Instead of commercial fire-lighters to light my fire, I use a small piece of paper and a good handful of pine needles and some pine cones and then put the wood on top of that. I've never failed to start a fire right away."

Sent in by: Maureen, California

"The droppings from our sheep and their leftover hay are used as fertiliser in our banana plantation."

Sent in by: Sukai Bojang,
Writer, Gambia

"Egg shells and egg boxes are really good for the compost."

Tip 96

"I use up old perfume or body spray as a room freshener. You can also give it to the recycling collectors. They use it to counter any pong from the food waste collections."

Mmmah, that's better

Tip 97

"Save Christmas cards to make next year's gift tags and place cards. Bigger cards can even be made into small gift boxes. Simply trace a cube template onto the card, cut it, fold it into shape and hey presto you have a gift box."

Sent in by: Kate,
Founder of green pages website,
www.greenfinder.co.uk

"Go green and get creative this Christmas by wrapping your presents in reused paper such as newspaper, wallpaper and magazine or old book pages taped together. Create coordinating bows or rosettes to make your parcel look pretty."

Sent in by: Anna, Hampshire

"At Christmas I collect up all the gift bags, bows and ribbons to reuse which saves a fortune! When I do have to buy wrapping paper I make sure it is 100% recycled and 100% recyclable."

Tip 100

Beno the Barba

"I save various bits and bobs that aren't collected for recycling and give them to my local scrap store. It collects all kinds of things like paint, CDs, and buttons which are then used as art and craft material for local play groups and schools."

What's your tip?

www.gizmo-the-geek.net

Things you can do with this book when you have finished with it!

- You can display it on your coffee table, left open on your favourite tip.

- You can return it to the author, who will reuse it. Please check the publisher's website for address details:

   www.greenlanespublishing.com

- You can take it to your local charity shop.

- You can give it to a friend, wrapped up in newspaper tied with the ribbon that came on your Primark pyjamas.

- You can leave it next to your kettle, open at Tip 27.

- You can leave it in your downstairs loo in the hope that it might interest passers-by when they have a moment or two to contemplate.

- You can share it with your family in the hope that you might save some money for a few treats. But watch out for page 56 if you don't already help out with the laundry.

Recommended Reading

How Bad Are Bananas? The Carbon Footprint of Everything by Mike Berners-Lee

Waste: Uncovering the Global Food Scandal by Tristam Stuart

The Story of Stuff by Annie Leonard

Reduce, Reuse, Recycle: An Easy Household Guide by Nicky Scott

To order any of these books visit:

www.greenlanespublishing.com

Bibliography

Berners-Lee, M. (2010) *How Bad Are Bananas? The Carbon Footprint of Everything*, Profile Books

Biddle, M. (2002) *Fact Sheet: Recycling the Hard Stuff*, Environmental Protection Agency, online at: http://www.epa.gov/osw/nonhaz/municipal/pubs/ghg/f02023.pdf

Bull, J. (2012) *Embodied Energy and Embodied Carbon of Water*, blog post, online at: oco-carbon.com/2012/03/15/embodied-carbon-of-water-update/

Carbon Trust (2011) *Conversion Factors: Energy and Carbon Conversions*, August 2011 Update

DECC (2011) *National Renewables Statistics*, online at: https://restats.decc.gov.uk/cms/national-renewables-statistics/#key

DECC (2012) *Renewable Energy in 2011*, online at: http://www.decc.gov.uk/assets/decc/11/stats/publications/energy-trends/articles/5629-renewable-energy-2011-et-article.pdf

DEFRA (2011) *Anaerobic Digestion Strategy and Action Plan: A commitment to increasing energy from waste through Anaerobic Digestion*, online at: www.defra.gov.uk/publications/files/anaerobic-digestion-strat-action-plan.pdf

DEFRA (2011) *Domestic Water Saving*, online at: www.defra.gov.uk/environment/quality/water/conservation/domestic

DEFRA, *Waste and Recycling*, online at: http://www.defra.gov.uk/environment/waste/

DEFRA (2012) *Waste Hierarchy Guidance Review 2012*, online at: http://www.defra.gov.uk/environment/waste/legislation/waste-hierarchy/

Environment Agency (nd) *Quantifying Carbon Emissions from Water Use in the Home*, online at: http://www.environment-agency.gov.uk/business/topics/water/109835.aspx

Environment Agency (nd) *Saving Water*, online at: http://www.

environment-agency.gov.uk/homeandleisure/beinggreen/117266.
aspx

Environmental Protection Agency (2011) *Greenhouse Gas Equivalencies Calculator* online at: http://www.epa.gov/cleanenergy/energy-resources/calculator.html

European Commission (2010) *Report From The Commission To The European Parliament, The Council, The European Economic And Social Committee And The Committee Of The Regions on the Thematic Strategy on the Prevention and Recycling of Waste*, online at: http://ec.europa.eu/environment/waste/pdf/Progess%20report.pdf

Ewall, M. (nd) *Primer on Landfill Gas as "Green" Energy*, online at: www.energyjustice.net/lfg

Friends of the Earth (2000) *Greenhouse Gases and Waste Management Options*, online at: www.foe.co.uk/resource/briefings/greenhouse_gases.pdf

Hudson, A. (nd) *Can Libraries Survive in a Digital World*, BBC, online at: http://news.bbc.co.uk/1/hi/programmes/click_online/9421996.stm

Hutchinson, A. (2008) *Recycling By the Numbers: The Truth About Recycling, Popular Science*, online at: http://www.popularmechanics.com/science/environment/recycling/4291576

ITV Meridian (2012) *The Hosepipe Ban and You: How to Save Water?* online at: www.itv.com/news/meridian/update/2012-04-05/how-to-save-water/

Kazmeyer, M. (nd) *How Much Energy Does Recycling Save?* National Geographic

Kemp, M and Wexler, J. (2010), *Zero Carbon Britain 2030*; A New Energy Strategy: The Second Report of the Zero Carbon Britain Project, CAT Publications

National Geographic, *The Hidden Water We Use: How Much H_2O is Embedded in Everyday Life?*, online at: environment.nationalgeographic.com/environment/freshwater/embedded-water

Nuffield Council on Bioethics (2011) *Biofuels Ethical Issues: A Guide to the Report*, available for download at www.nuffieldbioethics.org

Pimentel, D. & Patzek, T. (2005) *Ethanol Production Using Corn, Switchgrass and Wood; Biodiesel Production Using Soybean and Sunflower*, online at: http://www.c4aqe.org/Economics_of_Ethanol/ethanol.2005.pdf

Planet Science (nd) *Powered by Chip Fat: the Cars of the Future?*, online at: http://www.planet-science.com/categories/over-11s/technology/2010/12/powered-by-chip-fat-the-cars-of-the-future.aspx

Science and Development Network (2010) *Are New Biofuels the Ethical Answer*, (blog post) online at: http://www.scidev.net/en/opinions/are-new-biofuels-the-ethical-answer-.html

Scottish Environment Protection Agency (nd) *The Waste Hierarchy*, online at: http://www.sepa.org.uk/waste/moving_towards_zero_waste/waste_hierarchy.aspx

Stockholm International Water Institute, *Statistics*, online at: siwi.org

Stuart, T. (2009) *Waste: Uncovering the Global Food Scandal*, Penguin Books

Unilever (2011) *UK Sustainable Shower Study*, online at: http://www.unilever.co.uk/aboutus/newsandmedia/2011/sustainableshowerstudy.aspx

UK Government (nd), *Greener Fuels for Cars*, online at: http://www.direct.gov.uk/en/Environmentandgreenerliving/Greenertravel/Greenercarsanddriving/DG_191580

Water Wise, *Fun Facts*, online at: www.waterwise.org.uk

Waste Aware (2009) *Can You Calculate the CO_2 of Waste?* online at: http://wasteawarebusiness.wordpress.com/2009/03/16/can-you-calculate-the-co2-of-waste/

WRAP (2011) *The water and carbon footprint of household food and drink waste in the UK,* online at: http://www.wrap.org.uk/content/water-and-carbon-footprint-household-food-and-drink-waste-uk-1

A note about the print versions of this book

There are two print versions of this book.

One version is printed on recycled paper in short print runs when we know how many books we need and that is enough for a viable print run. The other version is printed on FSC certified paper by a print-on-demand printer.

At the time of printing this edition I have been unable to find a print-on-demand printer using recycled paper. Print on demand is green in itself because it avoids using up resources for no reason. The books are only printed when someone wants to buy one and they are shipped directly to that person, avoiding any overproduction, reducing miles the book travels and removing the need for warehousing.

I would prefer to be able to print all copies on recycled paper but that doesn't seem possible at the present time. Things may change, but for now, I hope, I have found the greenest solution by using a combination of technologies.

Gizmo the Geek

Who is this Gizmo the Geek character who is kindly holding up the number of each tip?

Gizmo the Geek and friends started life as three short stories about rubbish and recycling.

Soon after, **gizmo-the-geek.net** was created as a place to store and share resources for young people and to collect and share 'ecotips' - little things we can all do for free that collectively make a big difference.

101 Ways is the first book in the Gizmo the Geek range. Further titles will follow as my research into turning rubbish into resources continues.

Readers' Views

As my book started to take shape and various drafts were circulated it became quite a talking point. It seems there is a little green core in all of us just waiting to speak out. After reading this book so many readers have said that they have a tip or two to share.

Please do come along and share your tips at either or both of my blogs:

http://rosiesecoblog.blogspot.co.uk and

http://gizmo-the-geek.net

There is a forum available for reviews, on-going discussion, for sharing tips and further resources on the Green Lanes Publishing website.

www.greenlanespublishing.com

Before you pass on this book to someone else, or return it to the author to pass on, you may like to add your name at the back of the book and say which is your favourite tip. You could add a tip of your own too.

Acknowledgements

A big thank you to everyone who has sent in an eco-tip for inclusion in this book or on the blog.

Thanks to all my fellow CWNM-ers, several of whom are contributors. Big thanks to The Florio Writing Group for their help, support and critique. Thanks must also go to Arvind Devalia and Stephanie Hale for their advice and encouragement and for keeping me on the straight and narrow whilst writing this book.

Thank you to Pure Recycling, Good Energy, Nick Cliffe at Closed Loop Recycling, Nick Peacock at Ecoplastics, Richard Garfield and Victoria Hutchin from Kier, Doug Teesdale and his team at May Gurney, and West Oxfordshire District Council for helping me with my research and giving me their expert opinion on my text. A big thank you to Cathy Wybrants for her advice with the science and Julian Easterbrook for his help with the maths. Thank you to Sarah Ellison for her insightful comments and encouragement, as well as for giving me the benefit of her excellent networking and marketing skills. Thank you to my various editors and reviewers. It takes many people to create a book.

Other than help with research, writing support and assistance with the myriad of tasks involved in producing a book, the inspiration behind that book is also of paramount importance. This I owe to my Mum and Dad for helping me to appreciate the wonders of this planet and for introducing me to The Centre for Alternative Technology in Machynlleth, Wales, where as a frequent visitor and supporter I have learnt so much and felt so inspired. Mum, I still miss you but I'm still driven by you, by the amazing energy, determination and competence you showed throughout your too short life. Dad, you are a rock, always there to help with random requests for technical guidance and research. And I'd never have done this without your support.

Likewise, I am ever grateful for the love, support and encouragement of my husband and children. To Richard, for reminding me I'm not doing this on my own, we are in it together, and to Lizzi and Jen for their practical help as well as for encouraging me to follow my dreams as I hope I encourage them to follow theirs.

Thank you to all those travelling with me on the journey of finding little things that make a difference, embracing leftover pie and putting our dustbins on a diet.

About the Author

Anna Pitt is a systems analyst, writer and blogger with a keen interest in green issues. She writes two eco-themed blogs:

http://gizmo-the-geek.net

and

http://rosiesecoblog.blogspot.co.uk

Her digital fiction work, *The O2 Tales*, was shortlisted for the inaugural New Media Writers' Prize at Poole Literary Festival in 2010.

Anna's family and friends manage to put up with her insistence on living without a rubbish bin. No one who knows her dares ask about throwing anything away. She provides waste-free cricket teas for her local village team and she makes yummy fridge cake. (Her lemon drizzle cake is not bad either).

Anna has a B.A. (Hons) in French and Education from Froebel College in London, a Master of Arts in Creative Writing and New Media from De Montfort University in Leicester and she holds the City & Guilds 7306 Further and Adult Teachers' Certificate.

Anna's 'Waste Not' and 'Dustbin Diet' workshops can be booked by councils, schools and community groups via the website.

www.greenlanespublishing.com

About the Illustrator

Toni Le Busque trained in Fine Art around the same time the internet was being thought up.

Her creativity flows into many fields, a bit like a broken hosepipe.

She has a B.A. and post grad in Painting, though she actually used pastels instead of oils.

She also has an M.A. in Creative Writing and New Media, though she doesn't Tweet.

She has a studio in Oxfordshire where she draws and also tattoos.

She makes a mean spinach risotto, substituting pearl barley for Arborio rice.

Her work can be seen at www.tonilebusque.com.

Her email is lebusque@icloud.com.

And she'll give you her phone number.

If you ask politely.

This book has been read by:

Name	Favourite Tip	Reader's Tip